PRESSURE-FRACTURED ICE, LAKE McDONALD. The awe-inspiring landscape of Glacier National Park—sharply chiseled peaks and broad, deep valleys filled with lakes, streams, and waterfalls—is the result of 3-million years of glaciation. Successive waves of glacial ice, in some valleys up to 3,000 feet thick, have quarried and gouged this landscape. The last Ice Age ended some 12,000 years ago, having peaked around 10,000 years earlier. The 40 to 50 small glaciers found in the park today are the result of the Little Ice Age of 1300 to 1700 AD. All of the glaciers in the park, including Sperry, Grinnell, and Blackfoot, are all much smaller in size today than when first described in the early 1900s. Interestingly, the park was not named for the number of glaciers evident today, but rather for the landscape created by them in the past.

From the 'WISH YOU WERE HERE' POSTCARD BOOK—Waterton-Glacier

FIRST CLASS POSTAGE REQUIRED

SIERRA PRESS

PHOTO: ©CARR CLIFTON

ASPENS NEAR ST. MARY LAKE, AUTUMN. The breathtaking diversity of plants found in Glacier National Park (more than 1,200 species) is, at least partially, the result of geography. The great physical barrier of the Rocky Mountains, actually the Lewis Range within the park, is the eastern limit of plants typical of Oregon, Washington, and Idaho and the western limit of plants typical of southern Alberta and eastern Montana. In addition, it is the southern limit of plants found in the far north of Canada. Within the park's one million acres this blending of plant communities creates a unique ecosystem worthy of national park status.

From the WISH YOU WERE HERE... POSTCARD BOOK—Waterton/Glacier

FIRST CLASS POSTAGE REQUIRED

SIERRA PRESS

PHOTO: ©WILLARD CLAY

BEARHAT MOUNTAIN, HIDDEN LAKE, AND BEARGRASS.
Lying cradled in its glacial trough, Hidden Lake is bounded by
Bearhat Mountain (named for a Kutenai tribal leader) and the
western edge of the Continental Divide. A 3-mile round trip trail
(The Hidden Lake Nature Trail) leads from the Logan Pass Visitor
Center to this viewpoint. Beargrass, which only blooms every five
to seven years, is not actually a grass, but a member of the lily
family. It was named by Lewis and Clark in the early 1800s and
is a favorite browse of deer, elk, bighorn sheep, and squirrels.

From the WISH YOU WERE HERE™ POSTCARD BOOK—Waterton /Glacier

FIRST
CLASS
POSTAGE
REQUIRED

SIERRA PRESS

PHOTO: ©JON GNASS

LEWIS MONKEYFLOWER AT LOGAN PASS. The subalpine and alpine landscape at Logan Pass, at an elevation of 6,646 feet, is a wonderland of wildflowers during the short summer months—July to August. The Hidden Lake Nature Trail, a 3-mile round trip interpretive trail that is partially on raised boardwalks, leads through subalpine meadows lushly carpeted with colorful wild-flowers to a breathtaking view of Hidden Lake. The thin soil found at this elevation is easily disturbed and the raised trail protects the ground and fragile plants from the impact of thousands of hikers.

From the WISH YOU WERE HERE™ POSTCARD BOOK—Waterton/Glacier

SIERRA PRESS

PHOTO: ©LARRY ULRICH

FIRST CLASS POSTAGE REQUIRED

GRINNELL POINT AND SWIFTCURRENT LAKE. Rising above the waters of Swiftcurrent Lake, Grinnell Point separates the drainages of Grinnell and Swiftcurrent valleys. It is named for George Bird Grinnell, a turn-of-the-century writer and conservationist. His series of articles in *Forest and Stream* magazine and essay, "The Crown of the Continent," in *Century* magazine described the beauty of the Glacier area and put forward a plan to preserve it. Grinnell's efforts, and the support of Montana Congressman Charles N. Pray, led to the passage of legislation, signed by President William Howard Taft on May 11, 1910, that created Glacier National Park.

From the WISH YOU WERE HERE® POSTCARD BOOK—Waterton/Glacier

FIRST CLASS POSTAGE REQUIRED

SIERRA PRESS

PHOTO: ©MARKHAM JOHNSON

REYNOLDS MOUNTAIN AND CREEK. Reynolds Mountain is one of several peaks—including Clements Mountain, Mount Oberlin, and The Garden Wall—that vie for one's attention at Logan Pass. With an elevation of 9,125 feet, Reynolds Mountain rises some 2,500 feet above the subalpine meadows at its base. Because its crest is part of the Continental Divide, waters falling on the far side of the peak will make their way to the Pacific Ocean. Waters falling on the side seen here will flow north to Hudson Bay. Going-to-the-Sun Road, which crosses the Continental Divide at Logan Pass, is open only from mid-June to mid-October as up to 700 inches of snow may fall during the winter months at these elevations.

From the WISH YOU WERE HERE = POSTCARD BOOK—Waterton/Glacier

SIERRA PRESS

PHOTO: ©LARRY ULRICH

FIRST CLASS POSTAGE REQUIRED

LEWIS MONKEYFLOWER AND HEARTLEAF ARNICA, OBERLIN FALLS. Glacier National Park is justifiably famous for its wildflower displays. From the park's lowest elevation (roughly 3,000 feet near Lake McDonald) to its highest (Mount Cleveland, 10,466 feet) spring marches up the eastern and western slopes in a splendid progression of seasonal change. West of the Continental Divide fields of heather, gentian, beargrass, and glacier lily please the eye. East of the divide, where it is drier and colder, pasque flower, lupine, Indian paintbrush, gaillardia, aster, and shooting star put on their stunning displays. With advance planning, it is possible to experience spring for months by moving uphill with the advance of the seasons.

From the WISH YOU WERE HERE . POSTCARD BOOK—Waterton Glacier

FIRST CLASS POSTAGE REQUIRED

SIERRA PRESS

PHOTO: ©JACK DYKINGA

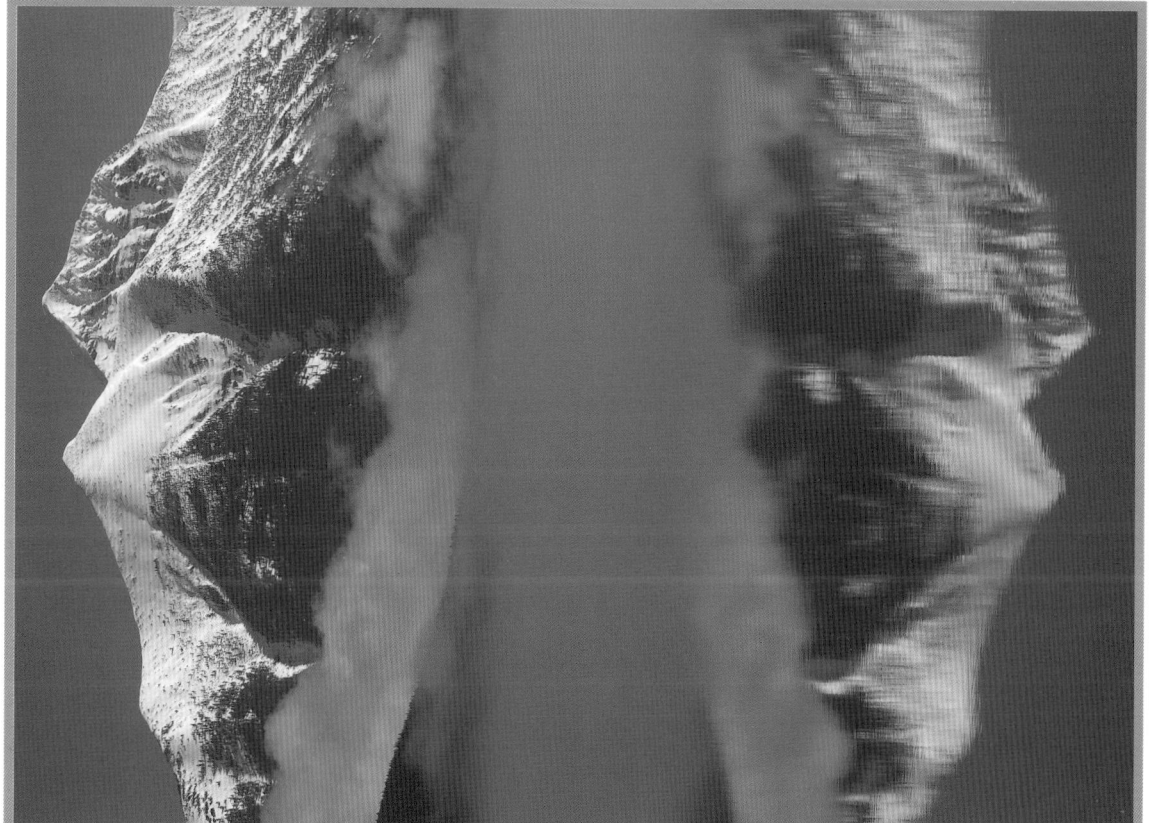

LAKE McDONALD. The largest lake in Glacier National Park, Lake McDonald is 10 miles long and 472 feet deep. Carved by a glacier that was 2,000 feet thick, it is known to the Kutenai as "sacred dancing lake" and was the site of sacred rituals conducted by these American Indians. Going-to-the-Sun Road parallels the southeast shore of the lake as it climbs toward the Continental Divide, towering in the distance.

From the WISH YOU WERE HERE... POSTCARD BOOK— Waterton/Glacier

FIRST CLASS POSTAGE REQUIRED

SIERRA PRESS

PHOTO: ©STEVE TERRILL

SUMMER SUNSET FROM NEAR LOGAN PASS. Within the one-million-acre embrace of Glacier National Park are grasslands, lush forests, and alpine tundra. Flora and fauna from the north, east, and west intermingle within this spectacular setting. More than 1,200 plant species shelter and sustain 63 native mammals, 272 birds, 5 amphibians, and nearly two dozen fish (not all of which are native.) Indeed, with the exception of bison and woodland caribou, every mammal native to the area is still present. This list includes, among others, black and grizzly bear, moose and elk, whitetail and mule deer, mountain goats and bighorn sheep, beaver and river otter, and wolves—who have returned to the park recently after not having been seen in Glacier since the 1930s.

From the WISH YOU WERE HERE POSTCARD BOOK—Waterton/Glacier

FIRST
CLASS
POSTAGE
REQUIRED

SIERRA PRESS

PHOTO: ©GALEN ROWELL/MOUNTAIN LIGHT

RAINBOW AND ST. MARY LAKE FROM SUN POINT. Here, near the eastern end of Going-to-the-Sun Road, is the place the Blackfeet Indians called "the lakes within." Going-to-the-Sun Mountain, which towers over the western end of the lake, was also named by the Blackfeet—according to their legend Napi, the creator, who had come to their aid, climbed the mountain in order to return to the sun. The Blackfeet Nation, who dominated the northeastern plains during the 18th and 19th centuries, was composed of three tribes: the Siksika (Northern), the Kainah (Blood), and the Pikuni (Piegan). Lands west of the Continental Divide were occupied by peoples of the Kutenai, Kalispel, and Salish cultures. Many place names in the park and surrounding landscape pay tribute to these early occupants who continue to live in the region.

From the WISH YOU WERE HERE... POSTCARD BOOK—Waterton/Glacier

FIRST CLASS POSTAGE REQUIRED

SIERRA PRESS

PHOTO: ©GALEN ROWELL/MOUNTAIN LIGHT

MOUNT OBERLIN AND BIRD WOMAN FALLS. Dominating the eastern end of the valley of McDonald Creek—the route of Going-to-the-Sun Road—Mount Oberlin rises to an elevation of 8,180 feet. Bird Woman Falls, near its base, is a tributary of McDonald Creek and also a classic "hanging valley." It was created when the glacier that carved the McDonald Creek drainage cut more deeply than the smaller glacier in the tributary valley. When the glaciers retreated, some 12,000 years ago, this small side valley was left hanging above McDonald Creek. Hanging valleys and broad, flat bottomed valleys are among the clues of a glaciated landscape recognized by scientists all over the world.

From the WISH YOU WERE HERE POSTCARD BOOK—Waterton/Glacier

FIRST CLASS POSTAGE REQUIRED

SIERRA PRESS

PHOTO: ©CARR CLIFTON

CHIEF MOUNTAIN AND GRASSLANDS. Named "Kings Mountain" by Peter Fidler, a Hudson Bay Company agent in 1792, Chief Mountain rises above the grasslands of the Great Plains in the northeast corner of the park. Its elevation (9,080 feet) makes it visible from a great distance and it was used as a landmark by the Lewis and Clark Expedition in 1806. The Blackfeet consider it a sacred site. Chief Mountain is the easternmost extension of the Lewis Overthrust, the great geologic event responsible for the formation of the ancestral Rocky Mountains in the region of Glacier National Park.

From the WISH YOU WERE HERE® POSTCARD BOOK—Waterton/Glacier

FIRST CLASS POSTAGE REQUIRED

SIERRA PRESS

PHOTO: ©CARR CLIFTON

ICEBERG LAKE, MANY GLACIER AREA. Although 3-million years of glaciation was responsible for creating the landscape of Waterton/Glacier—such as this glacial cirque complete with 3,000-foot headwall—the raw materials upon which the glaciers acted are more than one-billion years old. Originally deposited in a shallow sea as silt and sediment, they were later compressed and solidified by their own weight—becoming shale, sandstone, limestone, and siltstone. Later, some 60-million years ago, forces from within the earth's crust caused the layers to compress—which resulted in warping, folding, and fracturing. Ultimately, a great break in the surface occurred and a 300-mile long slab of the earth's crust (up to 15,000 feet thick) was lifted and thrust 50 miles to the east. This event, known as the Lewis Overthrust, was responsible for providing the raw materials for wind, water, ice, and glaciers to work on for the last 50-million years, creating the landscape we see today in Waterton and Glacier National Parks and the Rocky Mountains.

FIRST CLASS POSTAGE REQUIRED

From the WISH YOU WERE HERE POSTCARD BOOK—Waterton/Glacier

SIERRA PRESS

PHOTO: ©GEORGE WUERTHNER

WILD GOOSE ISLAND AND ST. MARY LAKE, SUNRISE. St. Mary Lake is one of the largest lakes within the 1,500-square-miles of Glacier National Park. The spectacular valley it occupies was carved by glaciers that reached their maximum size some 20,000 years ago. The lake (9 miles long and 289 feet deep) and its setting were known by the Blackfeet as "the lakes within." The eastern end of Going-to-the-Sun Road runs along the north shore of the lake.

FIRST
CLASS
POSTAGE
REQUIRED

From the WISH YOU WERE HERE, POSTCARD BOOK—Waterton/Glacier

SIERRA PRESS

PHOTO: ©WILLIAM NEILL